THE TIME PARTICLE

I0504127

A Christian theory of time
and things mysterious

Author

Terry McIntosh

THE TIME PARTICLE

DEDICATED

To people of faith seeking to understand the act of creation and things mysterious that Christians don't talk about.

Jeremiah 33:3:
"Call to Me, and I will answer you, and show you great and mighty things, which you do not know."

Contents

Introduction

This theory of time and things related is just that - a theory in layman's terms. In science, it's an unproven hypothesis. A particle is manifestation of a field that generates mass through its interaction with other particles. In context of identity, this particle is part of a larger complex system that exists in another realm. For sake of simplicity, I have named it the Time Particle.

This is a short read but loaded with – well, you will decide what it's loaded with. We live in a new age where knowledge abounds. New discoveries and advances are made regularly. Many of them appear contrary to the Christian belief in a triune God who created everything. Science offers answers to legitimate questions whether right or wrong. If Christians don't give the right answers, it can be expected that many of the upcoming generations will accept and adapt to the answers they are taught by science and the universities of wisdom no matter how wrong or how far from truth conclusions may stray. One might say that God imparts faith to believe and that's enough. True, God imparts faith but it's not always that simple. He also gave us a mind to think with. The Bible tells us to reason with God. That means he wants us to seek him out. Isaiah 1:18 says, "Come now, and let us reason together, saith the Lord."

We are served parts of truth by the worlds interpretation of mysteries, but all truth matters. Truth is truth. This theory applies to those without faith as well as those of faith.

The revelation of how time works is imparted for edification of the church first. Secondly, perhaps it will serve science as well. God knows the result but I don't.

I will be repetitious at times for the sake of clarity and to avoid getting lost in a maze of new information.

This revised edition includes a broader look at Satan's ultimate plan to usurp God's plan for humanity.

HAWKING GOT IT WRONG!

Cosmologist Stephen Hawking got it all wrong about time and the act of creation. He concluded, "You can't get to a time before the big bang because there was no time before the big bang. We have finally found something that doesn't have a cause because there was no time for a cause to exist in. For me, this means there is no possibility of a Creator because there is no time for a Creator to have existed. "

Mr. Hawking failed to recognize the creator in quantum laws that allowed the big bang to occur. Time is eternal. Without time, there would be no big bang. Everything that exists was born in time, lives in time, and dies in time. It existed before the big bang. The big bang occurred in time.

"The Time Particle" hypothesis explains the before, during, and after big bang event. It does not violate quantum rules. Other mysterious things like black holes are readily understood when they are embraced from the quantum perspective of all things cosmic in the realm of a time particle.

Everyone has faith in something or someone or themselves. Those without faith in a creator might quickly dismiss the idea of time before time and condemn it as less than a serious work. After all, it is pseudoscience at this point. However, Truth is truth no matter what one thinks about it, and quantum rules overrules what someone might prefer to think. Truth applies to those with faith and those without it. It would be a mistake to ignore this revelation because of an anti- religious bias. A closed mind should not expect to learn anything new. It takes more faith to believe that everything came from nothing than it does to believe in a higher intelligence we call God.

We want to know "Where did we come from? How did it happen? Where are we going?" Can we change the future? Can we change the past? What if some of these mysteries could be explained with one simple revelation of time, what it is, and how it works? What if questions about predestination and free will could be answered by this same revelation of time?

Time is the answer, but no one really knows what time is (until now, of course). Physics define time by its measurement: time is what a clock reads. It is a measurable quantity and, like length, mass, and charge, it's usually described as a fundamental quantity.

Duncan Watson, a Bachelor of Science graduate with an honors degree in Physics & Applied Mathematics from the University of Na-tel, Durban (1968) says, "No one knows what time is. It is so mysterious that we cannot even measure it. We measure the movement of a pointer across a clock face. We do not measure time itself."

Researchers haven't figured out how time or the universe really came to be. The observable universe can be thought of as a sphere that extends outward from any observation point, much like blowing up a balloon. It can also be argued that it's flat, but it's my guess, based upon the Time Particle and "Big Bang" theories, I have to say the Universe is somewhat spherical in shape even if it appears flat from a distance perspective. Everything within the Universe is nearly spherical, and that's because enough gravity in a mass causes this shape, while the rotation tends to slightly flatten out the perfect sphere. Regardless, what caused nothing to explode into everything in the shape of a balloon like sphere? It's nothing less than a miracle event by any approach. The researcher is standing inside the balloon scratching his head while trying to figure out how he got there.

Science explains it this way. The Big Bang theory generally accepts that everything was compressed in a region so infinitely small that we would never be able to comprehend it. It was billions of times smaller than an atom. The general consensus is that time was set to zero and did not exist. Nothing did.

Then something happened. Without cause, an unknown force exploded itself outward. Space and time burst on the scene instantly creating the universe as we know it. Gases and dust within the universe consolidated into visible matter and emerged everywhere at once. The Bang eventually spewed out billions of galaxies with millions of stars each.

The theory generally concludes that everything that exists was formed out of nothing. Matter was created at the point of expansion. If there is no matter, there is no space, and no time. At the initial stages of creation, a single super force ruled everything. Something else caused the force to split and a new force called gravity emerged. It is what shaped the cosmos. The ultimate shape was and remains perfect. It made it possible to create DNA and life as we know it.

Billions of galaxies spread out evenly after the initial bang with the same number in every direction. It was a great mystery as to what source was behind such a balancing act. Nobody in the field of study initially understood it. The act of expansion and balance has since been labeled as "Inflation." It says that an energy force caused expansion with a sudden act of acceleration.

A grain of sand is often used as an illustration. Imagine it swelling up as large as the Sun at the speed of light. That's "Inflation." Science is satisfied that gravitational waves are the energy source behind it. The waves stretch space time.

I've wondered about this and other related subjects like time travel ever since I was a young boy. I marveled at how size is relative and that our universe might not be any bigger than a pin head. I put it on the shelf and moved on. Now, in recent times, I revisited the subject with renewed interest due to an unsolicited dream impartation.

The dream exposed something unique to my limited knowledge. The object was a series of moving energy parts. I was not familiar with anything like it and possessed no learned reference to explain it. I awoke and drew it on paper to preserve memory of it. I knew it had to do with time, but what?

Science offers explanations without the benefit of a sovereign creator and this is where Mr. Hawking and many of his peers have missed the mark of excellence. The Time Particle is a faith response to an otherwise godless science. It resolves hard to answer questions by using the science that a sovereign God created.

The Big Bang theory and subsequent conclusions are drawn by looking at creation from within the creation and as a minor player within it. We've been able to snap images of stars and galaxies in the early stage of formation with the aid of the Hubble Telescope, but we can't get past the point of initial creation. We are on this side of the event. That's the wrong place to be standing.

In order to grasp the big picture, science needs a new concept of time. In his book, "Time Reborn," theorist Lee Smolin writes, "I no longer believe that time is unreal. In fact, I have swung to the opposite view. Not only is time real, but nothing we know or experience gets closer to the heart of nature than the reality of time...I believe that to make sense of the...universe, we must embrace the reality of time in a new way." Theorist Jim Baggott pitched in and said, "If time is real, space is an illusion."

I have not read Mr. Smolin's book, but I concur that time is real. How we approach it matters. We'll never see how things happened by peering backward into time. Imagine that we're driving our car toward the point of inflation in search of an explanation. We will always collide with an invisible "wall" at the point of creation and with less than desirable results.

It is generally believed that time did not exist before the Big Bang occurred, but when did the clock really start ticking? We must step outside of creation and witness it from the point of origin. In theory, we must stand with God on his side of all things and observe it from his higher plane of view.

This report is a compilation of truths I believe are absolute. Conclusions are rooted in quantum physics. If I'm right, it's all glory to God. If I'm wrong, the buck stops here. If I've drawn wrong conclusions, that's on me, not the Creator.

As an introduction to the creator, evidence declares that God is not a singular energy force or thing as some mystics suppose. He is a superior being far above our imagination and he was not alone in the beginning. Genesis 1:26 records that "God said, "Let us make man in Our image." So, God thinks, he strategizes, he feels, he has emotions. He made us in his (their) image. We resemble our Creator in our capacity to display the same attributes of love and other characteristics.

He delights in creating things and making his wonders known to us. Jeremiah 33:3 says, "Call to Me, and I will answer you, and show you great and mighty things, which you do not know." He is giving us a glimpse into his wonderful creation.
So, how did God do it?

SEE ILLUSTRATIONS A and B.

Illustration A:

Crude drawing displays energy motion with clockwise and counterclockwise directions splitting into two halves.

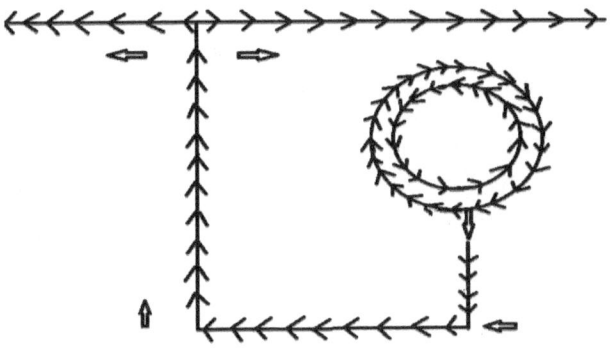

Illustration B: *Two Time Particles in original state are orbiting in opposite directions before expansion. The Pauli Exclusion Principle states that no two electrons can have the same quantum number, and thus, electrons in the same orbital must have opposite spins. So it is with the time particles.*

Electrons and time entanglement

We launch our investigation by going straight to the mystery world of Quantum Mechanics. That is the lowest known level of atomic action. It describes nature at the smallest scales of energy levels of atoms and subatomic particles.

Understanding the time particle is essential to understanding intangible time as we know it, and for making a case for divine creation. In context of identity, this particle is part of a larger complex system that exists in another unseen realm. There are things all around us that we can't hear or see with the naked eye to include color, light, and sound. Time is like that, but one doesn't have to be a Doctor of Physics or a professional theorist to understand it. We just need a few basic quantum rules to work with. They take us on an interesting and intense journey through the maze of a beautiful and awesome, wonderful creation.

The primary rule is this: If a physical event is not specifically forbidden by a quantum rule, then it can and will happen. This reflects

Christian faith in God's sovereign ability to do anything. In Luke 1: 37 the Bible says, "Nothing is impossible with God." We see that principle as real in the study of quantum physics. All things are possible. Adversely, the open door rule can lead a researcher into a fictional account of everything mysterious with means, ways, and action that God does not employ. Just because it is possible does not mean that's how it occurred when God performed it. The Time Particle leads us to a proper conclusion.

An all-important fundamental energy particle we need to understand is the Electron. It exists in the atom world and has mass. It has a negative charge and is a core element of ALL things. Everything that exists is comprised of electrons. That's you, me, the world, the stars, and everything. Physicists don't really know what electrons are. They know things about them and can observe how they act but they don't really know what they are made of. It's important to us because the time particle is intimately related to it.

Although an electron is not considered pure energy, it is a quantum object with wave-like properties. It is currently not known to be made of anything smaller than itself, but it is. I submit that the electron was born out of the time particle. In other words, the electron was created in time. Time is the prime particle that proceeds everything else. Since everything is created in time, it appears that time was the first thing God commissioned within his unlimited range of operation. Time is his workspace spread out like a director's story board or like an artist's blank canvas that was arranged for ardent, impassioned work.

Researchers are amazed to observe electrons popping up in one location, disappearing in another, all the while defying logic. Over time, time intersects with time, every time, all the time. Time bleeding into time can potentially explain the popping in and out of electrons.

Electrons can absorb energy from EXTERNAL sources and be promoted to higher energy levels. An example is photoexcitation, where the electron absorbs a photon and gains all its energy. When the electron falls back to the original ground state, the energy can subsequently be released in the form of a lower energy photon. Light is carried over space by photons. That explains what happened when God said, "Let there be light." Electrons released photon energies and wham-O, light appeared.

Electrons are known to consist of 3 parts. #(1). The Holon is the whole of it and a part that carries the charge. #(2). The Spinon carriers its spin and # (3) the Orbiton carries its orbital movement. When a single electron fractionalizes (splits) into 2 pulses, or two halves, the spinons and orbitons move in opposite directions. The particles can move with different speeds. Each half carries a particular part of the electron. The final state cannot be described as a single-particle state. It is, rather, a COLLECTIVE state composed of several energy levels. The God head is recognizable at the level of quantum physics. The electron collective state of two is part of the whole but are separate. They move independent of each other though they are part of the whole.

This basic quantum fact reflects the Christian belief in a triune God referred to as the Trinity – Three in One, Father, Son, and Holy Spirit. The whole represents God as One. The collective state of two represent the Son and Holy Spirit respectfully. Any one part can speed up or slow down at various points in the journey. They are all members of the same God family. The collective electron occupies multiple spaces at the same time. According to faith, God does, too.

The Pauli Exclusion Principle states that no two electrons can have the same quantum number, and thus, electrons in the same orbital must have opposite spins. So it is with the time particle.

Quantum Entanglement occurs when separated particles interact with each other. Changing the state of one will change the other but they remain separate and distinct even though it is one whole. Mind wobbling? At first, yes, but the mystery springs to life when we think it through. There are some things we just have to accept even at mind wobbling heights. It's just a fact. The fractionalized electron is one subatomic particle in two places at the same time. They are identical. What affects one affects the other. If you have one in your basement and its twin is on the moon, and if you affect change in the one on earth, it will also change the one on the moon.

Or imagine that you have an electron twin who lives on the other side of the world.

Whatever happens to you, happens to the twin. Whatever happens to the twin happens to you. If you turn right, your twin turns right. If he falls ill, you fall ill. That's exactly how the fractionalized electron interacts with its other self. They are separate but one and the same.

Dark matter and dark energy also play a major role in understanding time. Albert Einstein was the first person to realize that empty space isn't "nothing." It possesses its own energy and defies conventional logic.

DARK MATTER is the invisible fabric of space but totally not understood. It produces the attractive force called gravity and holds everything together. For those of us with faith, we believe that is through Christ. Colossians 1:17 says " He is before all things, and in him all things hold together." This theory suggests that dark matter existed BEFORE our atomic universe was created.

DARK ENERGY is another mystery. It is a property of space and lighter than gravity. It is an unknown form of energy which is hypothesized to permeate all of space. Both are necessary to make the universe function.

The space time energy plays a major role in expansion of the universe. Observations of very distant supernovae showed that, a long time ago, the universe was actually expanding more slowly than it is today. So the expansion of the universe has not been slowing due to gravity, as everyone originally thought that it would. Conversely, it has been accelerating. No one expected this, and no one knew how to explain it until advent of the Inflation theory which supposes that gravity waves are responsible for expansion. Although such waves contribute to expansion, the Big Bang theory says gravity originated after the bang. Gravity should have held the matter in closer proximity than what is happening. Expansion proves the existence of gravitational waves but that's not all there is to expansion. We have to ask "What propelled the energy to expand initially?" Nobody knew until now.

Dark matter is the invisible fabric of the universe within the Time Particle. It consists of unknown energy particles called sub-quantum energies. Nobody knows anything about those particles at date of this report. I suggest they existed BEFORE the big bang. In contrast to the Bang theory, there is no matter without time. Time permeates everything. Time Particle energy forced dark matter to explode outward generating energy motion to stretch and expand dark matter. Gravitational waves as described by the Inflation Theory plays a part after creation but there is more to it. Gravity came after the bang; therefore gravity waves were deployed secondary. A graviton is the hypothetical quantum of gravity, an elementary particle that mediates the force of gravity. It had been thought that the effect of gravitons would restrict continued expansion. However, it does not. Gravity hold objects in place at close quarter. Without it, we would float off the earth. It is not strong enough, however, to keep the universal neighborhood together.

The force of dark energy is lighter than gravity and keeps expanding. Time generated dark energy motion and stretched dark matter. It was imparted from an external source outside of the universe. It excited the atomic world and split into two halves like a fractionalized electron. An example would be to inflate a balloon. The air creates space within and expands the material. The air existed outside of the balloon before it entered the balloon. Air was forced into the balloon from a pre-existing, external source. You can't see it, smell it, or touch it, but it's there. Where did it come from? In the balloon example, it came from breathing air into the material by exhaling from outside of it.

The fabric of dark matter space existed before creation week. A small piece of Heaven spilled over when the Time Particle ejected it into the nothing and started the process of creation. This is where a traditional researcher may be tempted to stop because they can't imagine anything existing before the creation. Remember, he's still standing inside the balloon trying to figure out how he got there.

Something or someone unarguably forced something outward to form the universe. It is still expanding. The Time particle is the energy force behind it. It came from a source outside of our known atomic universe. That source is God. Hang onto that thought.

Two and three

Time is the answer to all things mysterious. It's an actual thing collaborating with the material of dark matter space that expands the universal boundary. Time is a blanket wrapped around everything. Our universe and everything in it was created by and formed in time. We are immersed in it. Time affects matter. It is the primary force behind expansion of the universe. It orbits the fabric and pervades it creating time waves thus causing continuous expansion. Without time, we would cease to exist.

We must understand time as it relates to creation before we can understand the whole of reality that include mysterious things like predestination, free will, dreams, visions, and other forms of prophetic knowings. So, in theory only, let's step across the great divide and stand beside God to see what actually happened. Consider this presentation His invitation to post witness the event.

Time as we acquired it began when God split "our" DNA designed Time Particle into 2 halves and they moved away from each other, thus creating TWO identical universes within our small atomic home. One supports the other. We exist in two places at the same time. Twin universes equals twin earths. What happens in one, happens in the other just as in the electron's fractionalized twin example. If possible to view the two universes in concert, they would consolidate into one particle in the view of the on looker. Two identical universes sharing space will appear as one universe to the human observer. That conclusion is rooted in quantum physics. The invisible twin was created at the precise moment when the Time Particle split and expanded dark matter. An invisible "umbilical cord" by any description, a tunnel, wormhole, or portal keeps both of them connected to the higher dimension point of origin.

As previously established, time co-existed in another dimension with dark matter. Time propelled it outward creating space and followed it, exciting dark energy. Time hastily completed one full orbit and dutifully created the universal boundary. The dual universe assures order and consistent expansion with excited pressure (applied energy) to the external boundary. The twin universe cannot leave the material in which they have been created, that being our atomic house. We can call the universal twins traversing time bands because of how they interact with each other. Bands stretch their respective universal boundary with consistent spinning that results in expansion of time and universal boundary.

We can envision it as a string of material pervading all of creation and orbiting around and in expelled matter holding the universe together.

It matters not which is Band 1 or 2 because they are identical. Numerical identification is for the purpose of illustration only. The time particle's DNA duplicates exact circumstances in both houses of the universe. They are identical bands of time and space. Time ignited the Big Bang. It originally moved in opposite direction at the same immeasurable split-second speed stimulating dark energy movement and stretched dark matter outward.

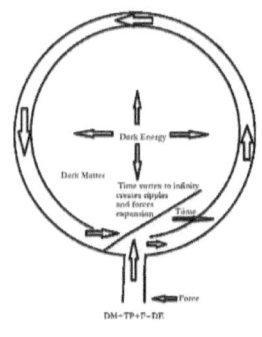

ILLUSTRATION C

It happened fast. It expanded like air in the balloon at an immeasurable speed. We should not apply our understanding of speed to the unknowns of the universe where physics break down. In Lawrence Krauss's book, he mentions that "for a very short period of time, so small it cannot be measured, an electron due to the uncertainty principle can appear to be moving faster than the speed of light; another way to interpret this is that it's moving back in time."[1] I haven't read his book, just quoting from it.

From God's vantage point, everything that exists started and ended at such speed that you and I would think it began and ended at the same exact moment. It was over in a flash. If we had stood next to him when the project was launched, we would have seen all of history unfold in one split second. Every human experience, every choice, every tear, failure, and every success would have played like a movie on big screen.

Initially expanded with maximum force velocity to the point of resistance, the bands slightly relaxed. They moved in curved directions determined by the perfect amount of force generated by the initial force at separation. A portal to heaven does exist. The invisible "umbilical cord," or portal, keeps both bands connected to the creator on the other side at point of origin.

Imagine two time bands stretched out in a straight line. Band 1 is positioned on the left. It expanded rapidly and curved in a clockwise direction. Band 2 expanded in an opposite counterclockwise motion and intersected with

---

[1] "A Universe From Nothing"; page 62

Band 1. Both curved and spiraled to complete one orbit and returned to point of origin. From there it continued with a new vortex orbit. The boundary bulged outward north, south, east and west. The bands continued to spiral orbit after orbit all the while expanding space boundary by generating invisible waves within the framework of dark matter in each band.

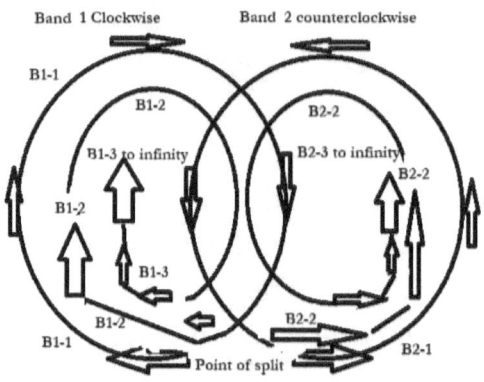

Illustration D: The first three orbits in the dual universe demonstrate how they intersect. Time bands swirl in opposite directions in close proximity. Space is quantum scale.

Orbits are tight and swing in at every potential point of contact to avoid collision with itself. Energy ripples generated by the vortex push dark matter outward expanding the Band (universe) in each one to accommodate more time. The bands move independently and float within limitation of external energies, thereby intersecting billions of times in micro-seconds. Time bleeds into time.

The two bands will consolidate into one at close of God's plan for the temporal home of humanity. It was an invisible force at work. Hebrews 11: 3 says, "By faith we understand that the universe was formed at God's command, so that what is seen was not made out of what was visible." We can say that a part of heaven was breathed into our universe.

Spiraling time bands within the boundaries of the dual universe we live in are forever orbiting and stretching while avoiding collision millions of times per second. Energy ripples exert pressure against the universal "wall" of each band, forcing dark matter to expand further. The swift, powerful, swirling vortex motion is what generates an undetectable wave of dark energy that stretch the dark matter. Example: A rock is thrown into water. The impact generates ripples in the water that move away from the point of contact. The ripples eventually lose energy and cease unless one keeps tossing rocks into the water to generate more ripples.

Incoming tidal waves serve as a grander example. Waves crash the shoreline and extends the boundary. Likewise, the energy force generated by swirling time usurps gravity and crashes the outer boundary of dark matter. Gravity was created in time, and denser than time itself in our time band. If time relaxed and stopped spiraling at immeasurable speed, it would cease to create ripples and Time would stand still. Gravity's DNA would assume divinely appointed control and essentially eat its parent predecessor and pull our universe back upon itself. The return to original state can only happen if Time Bands return to ground rest and Time runs out of room. Fortunately, continued expansion is triggered by internal pressure exerted from the time bands pushing against the universal boundary, thereby making room for more Time Band orbits and continued intersections of time.

Time bands run parallel and can cross each other at a near miss distance that does not disturb the trajectory of either one. The result is stable, consistent expansion of each universe. Order was arranged out of chaos. Time as a "thing" propelled a high energy force to stretch and expand the universe for the sake of survival. Expansion is speeding up as the bands orbit. It pushes or pulls planets, solar systems, and all things universal further away from each other. Accelerated speed indicates that time bands are generating more ripples in dark matter. That means the point of origin gets further and further away from the present.

A pessimist might say that the past is slipping away faster than ever. An optimist might say that tomorrow will be here before you know it.

We are not aware of external increasing time speed because it is relative. We live within the Time Band but are only aware of our material world which is governed by a set of internal laws of physics. We shouldn't confuse the speed of universal time bands with the rate of speed in our visible universe or that of planet earth.

We live in three time zones. When Band 1 intersects with Band 2 it will do so at a different point in time every time creating multiple time "zone" crossings. The bands are wound tight and so close to each other that the human eye would only see one of them. They cross each other at points that otherwise would result in a collision. Both bands are real timelines and those living at every point are very real with a past, present, and future.

While one time loop contains the present, future, and past, each moment is as real as the other. The B-theory of time is the name given to one position regarding philosophy of time. B-theorist students argue that the flow of time is an illusion, that the past, present and future are equally real, and that time is tenseless. I don't know how they came to that conclusion, but it's true according to the Time Particle Theory.

It is always the present for you right now although you are alive in the past and future. It is the present for your past self. It is the present for your future self. Example: You are reading this hypothesis. Your future self has already reviewed it in the past. Your past self will do it in the future. Each point in time is real. It is all you at various stages of Time Band stretching. You are in 3 different time zones at the same time.

As difficult as it may be to process, we are reminded of the three fold nature of the electron and the time particle. We are also reminded of the triune God three in one. And remember that with God all things are possible.

Time is God's memory book. It keeps us living each moment before him. It records and preserves all things. Hebrews 4: 23 says in part "Nothing in all creation is hidden from God's sight." He sees all past, present, and future. There is no end to what he knows. Every thought and every detail of our lives serve as a living testimony of everything we have ever done, said, and tried to hide. Our secrets lay bare before a God who can review every detail at any given moment in time. Imagine replaying a favorite movie and watching it over and over. That's disturbing on a personal level but God has promised not to remember our sins once we accept Christ. It is a divine choice to forget. He chooses to remember that we believed his testimony about Jesus.

We discover further support for timeless time in the Holy Bible. Psalm 139:4 says "Before a word is on my tongue you, Lord, know it completely." Isaiah 46: 10 says "I make known the end from the beginning, from ancient times, what is still to come. I say, 'My purpose will stand, and I will do all that I please."

The revelation of past, present, and future living encourages a secondary rendering of scriptural context. We have to read it with new understanding in context of the time particle. Luke 20:38 says "For He is not the God of the dead but of the living, for all live to Him." Ecclesiastes 1: 9-10 says "What has been will be again, what has been done will be done again; there is nothing new under the sun. Is there anything of which one can say, "Look! This is something new?" It was here already, long ago; it was here before our time."

The prophet goes on to say in Ecclesiastes 3:15 "There is a time for everything, and a season for every activity under the heavens: Whatever is has already been, and what will be has been before and God will call the past to account."

God sees us as alive past, present, and future. Nothing can escape close, holy scrutiny. When he looks at us in our present timeline, we are alive. When he looks at us in the past, we are alive. When he looks at us in the future, we are alive. We can't hide from God's eye past, present, or future. Time bears witness for or against us.

Now that we're familiar with structure of the Time Particle, we can move on to more fun things like "How does it affect life? Is it predestination or free will? How does prayer work and does it really make a difference? Can we change the future? Can we change the past? The next episode is an exciting adventure through a beautiful maze of time entanglement. Answers long sought are no longer elusive, so buckle up for the ride!

Predestination or free will?

If the Time Particle stopped expanding, the universe would follow gravity. Progressive time would stop. The bands would reverse intersecting over and over to the lowest point until there was no place to go. Everything would be consumed by a massive black hole of pure energy when the time particles collided at point of origin. Fortunately, it will not end that way. How do we know that? The Creator has seen the future and revealed parts of it to us. He knows. But how does he know? Did he script it? Is everything planned? Predestination says, "Everything that will happen has happened." But wait a minute! Is the future pure fatalism without free will choice, or is it the result of choices we have made, are making, and will make?

God observed humanity's journey to the very end. The human experience has concluded. The future has already happened and we are playing catch up in our present timeline. Our past is moving to it's future which is our present. Our present is moving toward our future at the finish line. Our omnipotent and omnipresent God moves into his past, which is also his present, where he engages our present and past lives. He shares information with us via means of communication and direct entanglement. He observes us at every station of existence. He stands ready to alter his past in response to our prayers and petitions. He erases information and straightens our approach which alters our future selves in rest at the finish line.

If this concept poses a stumbling block to reality, remember quantum rules. God made those rules and they reflect his design. He interacted with humanity along the way to lead us to a specific close of history. It was Duncan Watson who said, "If you believe in prophecy then you will realize that the future is out there somewhere in the realm of thoughts. But the only mind big enough to contain the entire future of the human race is the mind of God. He alone knows what is going to happen. He is an invaluable contact if you want to secure your future."

The original timeline ran its course in a fraction of seconds from the eternal viewpoint. We made choices that effected our lives and our eternal destination. Although he predestined some in advance for glory in the purpose of knowing Him, God allowed us to choose or reject him. Does this line of thought create a paradox? No. The triune God CHOSE not to reign over a world of robots. He wanted to interact with his creation so he breathed spirit into mankind. Genesis 2: "Then the Lord God formed a man from the dust of the ground and breathed into his nostrils the breath of life, and the man became a living being." The breath, or spirit, of God turned man from a lifeless collection of matter into a living creature. Without spirit, man is just a biological robot under orders. That is not what God wanted. He desired a personal, intimate relationship with his creation. He wanted us to choose him and interact with him in that love. Therefore, he gave us free will so that we could choose to love him or reject him.

But, then again, what about Ephesians 1:5? It says, "he predestined us for adoption to sonship through Jesus Christ, in accordance with his pleasure and will." Does the Time Particle theory contradict the Bible? No. He created some as objects of purpose and others as free will creatures. He created the Apostles for purpose and opened the door to others by choice. Ephesians 1: 13-14 says " And you also were included in Christ when you heard the message of truth, the gospel of your salvation. When you believed, you were marked in him with a seal, the promised Holy Spirit."

Now, God peers back into time and sees us alive and in process of making those choices that determined our eternity. He knows what choice we shall make because we've already made it. It stands to reason that if we are with Christ at the end, that we won't make a different choice in regard to accepting Christ and the plan of salvation. Therefore eternal security is guaranteed for those who come to faith in Christ.

God works in our lives but he lives outside of our timeline in another dimension. Originally, mankind made individual choices. The tower of Babel serves up an example. In Genesis 11, God heard about what they were doing and came down to see it for himself. This may sound like God was no smarter than humanity. He didn't know the future. He had to hear about what was going on. He had to investigate and take action to protect his plan. He came down from somewhere. "Come, let us go down and confuse their language so they will not understand each other." He came through the portal that connects both Time Bands to the point of origin.

He did all those things by CHOICE. He chose to limit himself in relationship to humanity. He allowed man to make decisions and intervened when those decisions were untimely or too far out of range of plan. In the case of Babel, God said that nothing would be impossible for mankind to do if they were in unity of speech and deed. So, he scattered them to delay progress. A new tower of Babel is being built today by the progressive one world government concept. That's a different study but we shouldn't read over what God said too quickly. "Nothing they plan to do will be impossible for them."

God created temporal time by splitting particles and chose to engage with it as in the Tower of Babel example. He stepped into time, times past, present and future, to guide humanity this way or that way. He spoke through time to the prophets and eventually incarnated as a human to live among us for a while. Colossians 1: 15-20 "The Son is the image of the invisible God, the firstborn over all creation. For in him all things were created: things in heaven and on earth, visible and invisible, whether thrones or powers or rulers or authorities; all things have been created through him and for him. He is before all things, and in him all things hold together. And he is the head of the body, the church; he is the beginning and the firstborn from among the dead, so that in everything he might have the supremacy. For God was pleased to have all his fullness dwell in him, and through him to reconcile to himself all things, whether things on earth or things in heaven, by making peace through his blood, shed on the cross."

Time has peaked in God's kingdom. According to prophecy, it's all over as far as his plans are foretold. Our future selves are at home in his new heaven and new earth. We will catch up to it, oh boy, happy day! So, how is it that God can predict the future if we truly have free will and could potentially decide otherwise? It's because the past, present, and future has already happened. He is presently sending information back into our past through the prophets and the Holy Bible. It is God time traveling and speaking through time into the past. We are living in his past. We're alive at every point of time, past, present, and future.

The Apostle John was caught up into the spirit realm and experienced a divine entanglement with Jesus Christ. While in that state of being, he saw things future to his present timeline. Events included doing away with the gulf between the heavenly dimension and earth. In context, John witnessed a future event like other prophets who saw future things. How could God show John the future if he didn't pre-destine it? It's not so complicated in the world of quantum physics.

Long before John's encounter, God witnessed every choice we ever made, are making, and will make. He interjected himself to guide us and help us but he gave us free will to choose what direction we take. He knows our choices and our final destination; therefore he can predict the future. Once saved, always saved" can be applied in context of what God knows. He knows the choice we made when we made it, as we're making it, and even before we make it. This theory lends credibility to my position on God's Divine Will and his Permissible Will. We live in both houses, divine and permissible. Example: In a future time to my past, God intervened to set me on the path to discovery about how time works. He did that through a dream. It was divine guidance. It was my choice to pursue it or not. I chose to chase after it. Now when God looks at me in that past event, he knows I will embrace the challenge. He knows because he witnessed my choice in the original sequence of time. He saw me make the choice to pursue it before the present me made the choice. On one hand, it was free will. On the other hand, it was predestined. We live in both houses. If one wants to argue free will, he can do so correctly because it was a choice. If another wants to argue predestination, he can do so correctly because it had happened before. Either conclusion appears valid at one time or other because we see everything from where we stand in time. Realizing our final destination has already occurred lends credibility to both once saved, always saved and predestination. Further realizing that we made choices in the original timeline supports awareness of free will. Imagine being inside a prism that separates white light into a spectrum of colors. I found myself standing in the center of one during a dream state. I heard a voice say, "What you see depends on where you're standing."

Light and colors reflected throughout the prism but it appeared different at every acute angle. When I looked one way, I saw one thing. When I looked the other way, I saw a different expression of light and color. That occurs because when light passes through a prism the light bends. As a result, the different colors that make up white light become separated. This happens because each color has a particular wavelength and each wavelength bends at a different angle. What we see regarding free will and predestination depends on where we are standing in the timeline. That's how it works. The Quantum physics rule says it's so. It is an absolute truth. It will become more clear when we discuss the observer effect.

Whether we adopt predestination or free will as reality, it serves us well to recognize that prayer and choice does make a difference as we shall see in the next section.

Prayer and the Observer Effect

Time Particle conclusions regarding prayer and how it works are simply beautiful. The potential for rejection is real because it is so incredible. The Lord revealed its mystery structure and I wrote it down, but it sounded like a fairy tale. I had no supporting evidence from any source. I accepted it as factual but was concerned that I couldn't explain it from the quantum world view. It's one thing to make a claim and another to qualify it.

I proceeded because I trusted divine guidance. During the course of putting this hypothesis on paper, I was encouraged by the Lord's input at difficult crossroads. That's what happened with the subject of prayer and how it works. I complained to the Lord. I asked him to help me explain it properly. Then, one morning, with slight effort on my part, new information came to light. It explained the praying "fairy tale" from within the quantum world. It was exciting! I was able to include the information and insert it before publication.

The act of prayer is how people of faith communicate with God. It determines our past, present, and future circumstances. If predestination was the sum conclusion, it would not be necessary to pray. We would just drift along doing what we have been programmed to do. The act of praying would itself be a programmed event. If we do bad things that would be God causing us to do it because he scripted it. Jesus taught us to pray. It's not a vain exercise. It makes a difference. It's for our benefit. Matthew 19: 26 says "Jesus looked at them and said, "With man this is impossible, but with God all things are possible." Remember that WITH GOD ALL THINGS ARE POSSIBLE. He exists outside of our time. He interacts with us in numerous ways. Faith and quantum physics agree that if he wanted to change something past, present, or future for a different result, he can. That's why we seek him out with prayers and petitions.

The fairy tale explanation goes like this: You experience a dream, vision, or word of knowledge as the result of time entanglement. We will engage the subject of entanglement with greater detail in another section. Suffice to say it warns you of impending trouble. It has already happened to your future self but it

CAN be changed by prayer and/or prudent action. God is giving you a chance, and more often helping you, to change a future event by altering the future result. To do nothing allows the negative circumstance to prevail. It has already happened and you will catch up to it. To engage with the power of prayer can alter what has happened to your future self with a reset of events.

It applies to your present self changing your future, which changes your future person's present or past. It applies to your past person changing his or her future, which changes your present or your past. Most of us would change something we said or did in the past if we thought we could. Imagine sending a message back in time to warn ourselves about a particular incident. Would we listen and take proper action?

A reset is complex in that it rarely effects just one person or event. It requires God to change the past from his perspective. Hence, the future is not set in stone until this great journey is complete. The future can be changed with the exercise of free will, but from God's perspective, it is predetermined. He knows the decision you will make. He sees you at every point of time and permits you to make the choice and, therefore, make a difference. So, from his point of view, the future is set in stone because he has witnessed every decision made in every timeline including every reset.

Any reset affects numerous other events. Therefore, not every future event can be altered because of subsequent interaction of numerous other factors, circumstances, and people. God will not permit all of our petitions because of the impact it has on many other people and circumstances. That's when we must trust his judgment over our own. A prophetic warning can prepare you for coming trouble, but not necessarily alter it.

Satan wants to usurp history and prove that God is a liar. He has read the Book and is trying to alter the future as it has been revealed. That's because he knows it CAN be changed. He knows how it prophetically ended in the original timeline. His present and past self is trying to alter the end result. Fortunately, God won't allow him to succeed.

Considering that we live in three timelines at once, let's label the past as #1, the present as #2, and the future #3. If #2 experiences entanglement with #3's time band and is made aware of a future bad circumstance, #2 prays for a different outcome. God hears, and responds positively. When #2 catches up to #3's timeline, the impending trouble or doom never materializes. Prayer changed our future and our future's past. Researchers will identify it as the Quantum Eraser effect.

Some theorists will suggest that we create separate timelines by altering events that have already occurred. That is not the case. God let us choose in the original timeline and changes it at will. It is all within the same lineage of time.

The fairy tale revelation takes us to the exciting subject of entanglement where it violates causality. The effect can proceed cause and seemingly re-write the past. According to Jon at respectyourintellect.com, gravitational lensing proves that "how we choose to act today influences particles that left billions of years ago, showing that our present decisions can affect the past and that time plays no role in the results of this experiment."[2] He means that it matters not how long ago something happened.

---

[2] https://youtu.be/eZUukl8CFdY

Jon's report blew my mind. It was a hard wrap around the brain. I reviewed his explanation several times before I grasped the signifance of it. Particles changed how they acted dependent on how they were observed with the result of altering their course. Remember the entangled electron and time particle in a previous episode? Entangled particles are able to pass information instantaneously across any distance. What happens to one happens to the other. Jon goes on to say "It's also clearly able to retroactively change the result of its entangled partner even if the measurement is performed AFTER the signal photon has ended its journey…"

Wow! That's stunning! Observing photons altered their past course and destination. The conclusion is generally accepted by science. Arvin Ash said, "The act of measurement of a particle NOW affects its properties in the PAST. A decision made in the present can influence something in the past."[3] It explains the fairy tale paradox that I could not explain.

Jon goes on to explain the Delayed Choice Quantum Eraser Experiment in another video.

---

[3] Extracted 10.04.2019 from https://youtu.be/0ui9ovrQuKE

[4] I refer you to his work for the full report while I condense it into one sentence. "The experiment indicates that the past can be changed." It sounds crazy but explains what the Lord puzzled me with. The fairy tale turned out to be factual in the quantum world. On the heel of that amazing find, another source unexpectedly popped up to clarify the issue. It's called THE OBSERVER EFFECT.

Researchers at the Weizmann Institute of Science conducted a highly controlled experiment demonstrating how a beam of electrons are affected by the act of being observed. The experiment revealed that altering an electron in motion alters its partner even though it has already reached its destination.[5] The theoretical basis for this marvel was developed several years previous and the Weizmann test results confirmed that theory. This means that objects observed alter their behavior in response to the observer. A past course was also altered by simply

---

[4] Extracted 10.04.2019 from https://youtu.be/HvDnMcHiLfs
[5] Extracted 09.23.2019 ScienceDaily at link
https://www.sciencedaily.com/releases/1998/02/980227055
013.htm?fbclid=IwAR1I9-
MFJqLYnDB_EoPEcBc4L4wF8f9PJtglaw-
2uDuXgcLMvXqz87rRP6U

observing photons in the gravitational lensing example as explained in Jon's presentation. A precious gem of knowledge was learned from it all. "The greater amount of watching, the greater the observer's influence on what actually takes place."

So, how does this apply to prayer? It's valid, scientific evidence that an observer can alter the course of events, including the past, simply by observing it. I identified it as a "reset." While the example of my present self praying and altering a future event is related, none of us can actually observe human events and reset what has already happened in our past. There are reasons for that. But there is one who can. He is the GREAT OBSERVER. He is God. He is Jesus Christ. He is the Holy Spirit. He is three in one. He is omnipotent and omnipresent, able to be everywhere at once with a glance. He is able to peer into his past at various points of our journey and alter it at will.

WE PRAY. HE OBSERVES. We pray longer. He observes longer. Occultic teachings offer a weak revision of the model without God and demands personal concentration at levels none can hope to achieve.

To say that God has a past is complex because he can observe everything in his present state. Everything that happens past, present, and future to us happens in his presence. We can say that God is a real time traveler. Everything and everyone is alive to him at all times. When he visits the past, it is a present moment for him. He draws close to us in our past, present, and future which is his past and present. When we offer up prayers and petitions, it's a present moment for him. It is complicated and simpler to say that God looks at us in his past because it has already happened.

He observes us in every moment of our time band. Jeremiah 23:24 "Can a man hide himself in hiding places so I do not see him?" declares the Lord. "Do I not fill the heavens and the earth?" Psalm 34:15 "The eyes of the Lord are toward the righteous And His ears are open to their cry." The idea frightens some, but we should want God to look upon us with kindness, grace, and mercy.

When God answers positively to grant our requests, he changes what is in his past. His observance changes the course of our present circumstance and resets our future events. The

Quantum Eraser effect is employed. Information is erased and replaced with new data. God sees us as a finished product. We don't. We see ourselves stumbling along the way hoping for a good finish. Considering our present, we are making choices. From his high and lofty place, he sees the end result of our choices and every single reset that he permits in response to our prayers. The point is that PRAYER MAKES A DIFFERENCE. Matthew 7: 7-8 "Ask and it will be given to you; seek and you will find; knock and the door will be opened to you. For everyone who asks receives; the one who seeks finds; and to the one who knocks, the door will be opened."

Should our present self be altered because of time entanglement and answered prayer, the effect is that our present and future self experiences a change in our circumstances. We go to bed one night with memory of an incident. Information is erased. We awake next morning with a different memory, result, and circumstance. We would not be aware of the reset save the occasional Deja Vue sensation. Likewise, our present self can affect a reset in our future person through prayer. It can only happen at the pleasure of the Great Observer. We exist in his past and he can change it at will.

In the next section, we'll get familiar with time travel and other spooky things.

Spooky things: Time travel, Deja Vue, Astral travel, Ghosts, and Reincarnation

Time travel is real, however unlike various movie depictions. If a traveler leaves earth and returns 100 years later, he will not have aged as much as those he left behind. That's actual, physical time travel for the astronaut but It's not like popping in and out of various timelines at will. An object must travel faster than light or stop traveling in time altogether to experience a shift in time. That's the current problem – we can't do that, but we shouldn't totally rule it out as a future possibility because of what Jesus said. The first rule of quantum physics agrees- All things are possible.

Perhaps would-be time travelers should consider ways to stop in time and let the past catch up to themselves instead of plotting to travel forward or backward. Even if a real solution is eventually discovered, that doesn't mean that it will be permitted. The wormhole theory of connecting two spaces of time would work if it was possible to survive the journey. That's a fun concept but not a practical idea for human time travel. Another sci-fi fun option is to zip around the universe faster than light and catch up to light waves from the past. That way, we could observe past events without interaction. But, alas, reality sets in. It is not possible to physically travel into the past and change events. God will not permit mankind the luxury of bringing disorder into what he has put in order. He is the real time traveler who moves from his present into the past to effect events and change history.

He further communicates with us from the future through prophets, dreams, visions, and revelation of hidden knowledge. It also happens when time entangles making time travel a reality in the astral realm because of time convergence. The rotating bands cross paths back and forth without colliding and often "bleed" over into the other. They rotate as close as milli-inches apart. That puts the future and past close to the present. Imagine future events being as close as the end of your nose. Wow! If only we could see it! We can, when time bleeds into time.

Deja Vue, as mentioned in a previous episode, is an example of time entanglement. It is a familiar feeling having already experienced the present situation. While there is evidence that Deja Vue is probably linked to the temporal lobe of the brain, the "probably" leaves it uncertain, and certainly not always the case. Scientists often expound upon things they don't fully understand with explanations they can't prove themselves, and there can be more than one explanation for an occurrence.

When a present moment entangles with a future or past event, it can create the familiar feeling "I've done this before." And, too, when a person "taps" into the entangled future time loop, he or she can experience what is called a premonition of things to come. That happens in real time, dreams, visions, and a gut knowing. Most scientists might consider "precognitive" incidents as flukes. While not everyone has a conscious knowledge about it, the person with practical experience knows it is a real phenomenon. It is often recognized as clairvoyance. All such incidents are steeped in God's way of doing things, howbeit not all are divine encounters.

If the spirit within the human body goes lighter than gravity, one can access the spiritual dimension through the locomotion of astral travel. The astral plane is a higher level of frequency beyond the physical. It is the realm of dreams, spirits, and psychic phenomena. It is the space between swirling time bands. If we enter the realm while sleeping, we can experience prophetic dreams and insightful direction. It is also the place of the dead awaiting judgment. Demoniacs and other crackpots access the realm, too.

Otherwise, Mystics, spiritualists, various fortune tellers, and some religious people, including Christians, claim the ability to astutely traverse within the realm at will by what is known as "out-of-body" excursions. While most claims are likely fraudulent, accessing the plane is a potential. It's easy to dismiss what we don't know as foolishness but I have experienced it and it makes sense in context of the time particle. The astral plane is a real place just out of our conscious reach.

In order to access the plane, one must go lighter than gravity. The human spirit does that at the point of death and departs into the spiritual realm. For the believer, he or she is caught up to be with the Lord outside of our time band. John 11:25-26, documents "Jesus said to her, "I am the resurrection and the life. Whoever believes in me, though he die, yet shall he live, and everyone who lives and believes in me shall NEVER die." 2 Corinthians 5: verse 8 says in part, "We are confident, yes, well pleased rather to be absent from the body and to be present with the Lord."

Others are not so fortunate and are restricted to the spiritual realm within close range of our time bands to await final judgment. I use the term earth-bound spirit to describe one so trapped. The Catholic expression of faith calls it purgatory. Contact is possible with departed spirits via mediums who have mastered the art of communication. The Bible story in 1 Samuel 28 about King Saul and the witch of Endor is an example. The Bible warns us to stay away from familiar spirits. It opens the door to malicious deception. Devils gladly pretend to be a friendly, departed spirit, but will eventually try to lead a person from Christ, or at least destroy his credibility with others. The only safe way to traverse the astral plane is under canopy of the Holy Ghost, and then for the right purpose.

The Apostle John was caught up in the spirit as he worshipped. God showed him wonderful things. Revelation 1:10 " I was in the Spirit on the Lord's Day, and I heard behind me a loud voice, as of a trumpet." John wasn't dreaming. He wasn't imagining things. He was actually there, in the astral realm where he heard and saw awe-inspiring things.

When one goes lighter than gravity, the potential to visit different points in time is real because of the intersection of time loops. The astral traveler can witness things, people, and events present, past and future. The caution is that the sojourner may encounter malicious entities resulting in physical possession of the body while the traveler is absent. The result could be possession by another soul that is waiting judgment. We need Christ as our covering when, and if, we are ever "in the spirit." It is a real, literal place between time bands, and much more than an emotional state of mind. It is hell in wait of final judgment for the dead outside of Christ. It must be tormenting to realize that all hope is gone. Flames of fear, regret, and despair fan the pain of separation from God.

Departed earth bound spirits cannot access the portal to Heaven. Luke 16: 25:26, says in part "between us and you a great chasm has been set in place, so that those who want to go from here to you cannot, nor can anyone cross over from there to us." They cannot easily access our material world just as we find it difficult to access the spiritual plane. They are lost souls.

Visible interaction with the junctions of time can and does happen. Scores of witnesses claim to have seen ghosts and other apparitions such as civil war soldiers in Gettysburg. They appear out of nowhere and disappear into somewhere. That's an example of interaction between time loops. It is a real present moment for the "ghosts" bleeding over into our present. It is also possible to be seen by those in other time loops. As a child, I saw a female spirit standing in a room with her back to me. She wore what I described as old time shoes and a long dress. She turned and looked at me before disappearing. Was she a departed earth bound spirit, an angel, or did she live in another time zone that bled over into my timeline? If so, she was as bewildered at seeing me as I was at seeing her. I would have peeked into the past, and she would have peeked into the future.

It is possible to willfully connect with the entangled band, however difficult. Gravity holds us in place and it's the culprit that keeps us from easily interacting with it. When we die, our spirits do go lighter than gravity. Until then, we are generally confined within the physical body, but there are exceptions as clarified by the out-of-body experience.

Some people appear to have an open spiritual "door" by which they can access past, present, and future information. They are often recognized as psychics, mediums, spiritualists, astrologers, prophets, and witches. They are conduit receivers with access to data in the swirling time bands. Many in that category claim to speak with or for the spirits of the dead. They don't understand how it works, only that it works. They can be good people but badly deceived people. Genuinely skilled practitioners are not to be confused with frauds who make false claims for fame and profit.

What the spiritist lets in through the "door" makes all the difference. Malicious spirits from past, present, or future timelines can impersonate angels of light and deceive the receiver. Belief in reincarnation serves as an example. It's the result of interaction with a past timeline and can be specifically encouraged by malicious spirits. The receiver, whether by word, dream, or vision engages a past event of someone else and concludes it was a personal experience. I almost fell for it after a detailed dream vision whisked me away to visit a previous time era.

It was early 20th century. I walked the streets. I passed people, T-model cars, and houses. I watched a blacksmith working in his shop. I walked home and saw cars parked outside the house. It was alive and real. When I awoke, it felt like I had really been there. I considered that it could be a past memory played back in dream state. After an encounter with Jesus Christ imparted faith to believe God's testimony, I realized that I had only witnessed a moment in time that belonged to someone else. It was not my life. According to Hebrews 9:27, I am destined to die once and face judgment.

Entanglement with a past life was more accurately an astral time travel incident. In fact, the receiver of similar visions, words, or dreams only witness a past event and assimilate it as his or her own. A notable experience was reported by two women famously known as "The Mouberly-Jourdain incident" which occurred in Versailles, France, August 10,

1901.[6] They allegedly experienced a time slip, and saw Marie Antoinette and some other people in the time of the French Revolution. Time entanglement provides clarity to what actually happened. Time bled over into time. The vision appeared to be current. It felt and looked like real time travel. It was, nevertheless, in the sense of time bleeding into time and they only observed it much like how I experienced a journey to the early 20th century while sleeping.

It appears that time travel on demand is restricted to the Astral plane. The only way we can change anything is through prayer and that is future bound. The past can be observed from the spiritual plane when time bands entangle, but not changed.

The future can be altered, but it's only the Great Observer, God, who can change the past.

---

[6] Extracted 09.12.2019
https://en.wikipedia.org/wiki/Moberly%E2%80%93Jourdain_incident#CITEREFIremonger1975

Black Holes

The mysterious Black Hole can be easily understood in context of the Time Particle. A black hole is a place in space where gravity pulls so much that even light cannot get out. The gravity is so strong because matter has been squeezed into a tiny space. We can imagine a star ten times more massive than the Sun compressed into a sphere approximately the diameter of New York City. Because no light can get out, people can't see black holes. They are invisible. This is generally attributed to a dying star imploding inward. Scientists remain baffled by the wonderful mystery, and numerous theories revolve around it. No matter how they look at it, at some point they shake their head and say, "We don't know."

Some ask the question, "Do black holes really exist?" Yes, they do, but science will never fully understand it until they can accept the Time Particle theory as presented in this series. Not unlike other spooky things, there can be more than one explanation for an event. So it is with the black hole. We can accept the general conclusion one is created when a massive star reaches the end of its life and implodes, collapsing in on itself. Otherwise, time itself creates black holes and insures order. We unwrap the mystery this way. Time is not constant within time bands. Environments with less gravity move slower and astronauts age slower than people of earth. Science agrees that the higher the gravity of a planet or star, and the closer to that body, the slower the time. Dual time bands, the twin universe, can travel at different speeds. It's DNA structure demands that what happens in one, happens in the other. When one speeds up faster than it's twin, it must slow down to allow the twin to catch up. Within the space of dark matter, time loops can create a tiny invisible bubble due to the immense speed generated when there is a need to slow down. Imagine slamming on the brakes and your car skidding around in a circle. It would create a time loop in dark matter. That's what happens when time needs to wait for its twin to catch up. The approaching twin duplicates the skid to slow itself down and stay in sync with the other slowing twin. Otherwise, it would blast on past its twin. This structure keeps the twins in constant sync and preserves order. The process repeats itself when one or the other speeds up again. Unlike single electron partners, they will eventually arrive to the final destination at the same time.

Due to the immense force generated by looping time, the tiny bubble consolidates into a single time bubble. It is much weaker compared to the whole of the universe. It is time within time and functions within the larger, universal band. It commences to exist within the boundary of its geographical reach and spirals outward. Space travelers passing through it would experience a shift in time. The mini time loop can be as small as a particle or as large as the imagination allows. The junior time band is real time in its own right, however contained within our original time band. It is single, smaller and less powerful but the same principle applies.

It is time within time. That is, a separate time within our time. If the junior timeline continued to expand it would eventually infringe upon the universe and result in chaos, but God will not permit disorder to disrupt what he has set in order. The expanding junior twin bands are weak compared to our primary bands. They eventually lose energy and return to the original state of ground rest. The junior, lifeless universe within our universe ceases to expand because it isn't strong enough to resist gravity. Gravity steps up to pull time inward and it reverses. The bands continue to spirally interact until forced to dock at the point of birth. This is where the rules break down and junior time stops. The process can be rapid or evolve slowly over millions or billions of years.

When expansion stops due to a relaxed state, a new rule applies and gravity takes over. Collapse is the only option left. Time folds back upon itself creating a black hole. Any travelers inside the junior loop would experience wild fluctuation in time but not be aware of the slowing down of time. It is all relative to the astronaut and the space time continuum. The event can't be observed in real time because time is invisible. It can only be observed after time stands still, thereby imploding. There are billions of black holes and more on the way. This is God's way of preserving order in our universe. The space traveler caught in a black hole will experience another major time shift as it spirals down, down, down, and down to the point of nothing.

I published this theory in the first edition on Nov 1, 2019. On Nov. 28, 2019, the Chinese Academy of Sciences announced discovery of a massive black hole that, according to them, 'should not even exist.' The newly discovered black hole, named LB-1, is located 15,000 light-years from Earth, according to a press release from the Academy. Scientists said the discovery would force them to "re-examine our models of how stellar-mass black holes form."

Was the timing of their discovery coincidence or divine revelation in support of the Time Particle Theory? It will not be considered a serious conclusion by scientists because the time particle is only a hypothesis to them. For me, it's a reality.

We'll look at the crucifixion and what it means to be Omnipresent in the next section.

Crucifixion & Omnipresent

We conclude this subject by dispelling doubt regarding God's sacrifice on the cross and his powerful abilities. The Bible proclaims that Christ died once to absolve sin. Someone might suggest that he dies multiple times in the Time Particle theory because time repeats itself. That would be a wrong conclusion. There is only one timeline in which the crucifixion occurred. It is the same event replayed in the history book of time. It is not a new, or different crucifixion. You may want to re-watch a favorite movie that features a character being killed off. That you go back and see it again does not mean that it happened twice in the story. It was a one time event. So it is with the crucifixion. It is the exact same event in the twin universe. It was and remains a onetime occurrence.

Otherwise, being omnipotent means that God has unlimited power and can do anything. Omnipresent means that he is present everywhere at the same time. How is that possible? It sounds like another fairytale, but it's true.

I have presented the Time Particle Theory as I received it. While I drew some conclusions based on experience, I stayed true to the time particle principle. Now, I must explain what I think about God's particular abilities by reason alone. He has not instructed me on this subject, but it fits the theory.

It should be obvious to anyone that a creator has unlimited power over his creation. That power enables the creator to come, go, do, and adjust anyway he chooses. Otherwise, to be everywhere at once requires deeper probing. How is that possible? We may prefer to think of God as holding us in his hands and looking at us all at once, seeing everything at once, but it may not work that way. Since God is the ultimate time traveler, it's easy to imagine him stepping into time at anytime. It is not a hard thing to step out and back a second later. By repeating the action, stepping in, out, and back in, he is with a person all the time. This method allows him to be with others the same way and what turns out to be the same time. He is essentially dividing himself through time to be with multitudes of people at the same time. Someone might say, That would take forever. Well, what is time to a timeless God? It's my opinion that it may well work like that.

We have unlocked a few of the many mysteries concerning time and how it works. While I claim divine inspiration and guidance, I hope you will judge it according to the theory itself, not the messenger.

Revelation of the unknown particle may or may not influence science going forward, but no matter what science concludes, we have the assurance of a loving God who knows all that is going on in our lives.

Understanding the Time Particle enables us to better understand our Creator. Various fantasy illusions are eradicated and replaced with practical application. It paints the image of a creative God who went to uttermost extremes to create everything. Such detail demonstrates how much he cares about his garden of life. It demonstrates how much he loves you in a personal way. If you consider how he observes you, how he steps back into his past numerous times to help you make good decisions, and to help you through difficult times, to make right what went wrong in the future, we should be humbled. To know that the creator comes to us and lives with us, and how he steps in and out of time so he can be with you all the time is more than humbling. He must really love and place value on you. True, he has workers who assist him. We call them angels and messengers of God, but they take orders from him. Truly, he is a wonderful, generous God. The Father, the Son, the Holy Ghost. Three in One. The Son lived among us in the flesh to introduce the Father directly to humanity and made a safe way for us at the cross. The cross is the ultimate expression of his love for you, for me, and all who will call upon his name.

He has been gracious to let us glimpse into the quantum world where we can admire his handiwork. I am eternally thankful for it and hope that you are blessed by it, too.

Now, we move over to the area of religious faith and pure speculation for the fun of it.

DAY ONE OF CREATION

Genesis 1: "In the beginning, when God created the universe, the earth was formless and desolate. The raging ocean that covered everything was engulfed in total darkness, and the Spirit of God was moving over the water. Then God commanded, "Let there be light"— and light appeared. God was pleased with what he saw. Then he separated the light from the darkness, and he named the light "Day" and the darkness "Night." Evening passed and morning came—that was the first day."

God got to work on it and established internal order within the newly expanded time bands. Time as we know it was established dependent upon rotation of the planet, stars, and things celestial. Earth would be governed by those internal time band laws.

How long was the first day? We can believe it was a 24 hour period, or we can believe one day lasted 1,000 earth years based on 2 Peter 3:8 "But do not forget this one thing, dear friends: With the Lord a day is like a thousand years, and a thousand years are like a day."

Thirdly, we can believe the controversial conclusion that one day likely represents millions or billions of years, rather an event period instead of "day" time.

The 24 hour period is possible because the Hebrew word for "day" implies it was in context of an ordinary day. The problem with this choice is that the heavens that determine times, days, and seasons did not begin until day four. What determined the length of the first 3 days?

Regarding 1,000 years in 2 Peter chapter 3, the context has nothing to do with the days of creation. Also, it is not defining a day because it doesn't say "a day is a thousand years." The correct understanding is derived from the context of the verse. It says, "one day is like a thousand years." The word 'like' demonstrates that it is a figure of speech instead of literal time.

The third option of a period of time stretching into millions of years is believable, too. But it matters not how long it took for the first day and subsequent days to elapse. No matter how long a time interval is from man's time-bound perspective, it's like the twinkling of an eye from God's eternal perspective.

It is an interesting question but not relative to the how things happened. Whatever one concludes is okay since it does not alter the process of creation and has nothing to do with salvation. Understanding the mysteries of God is not a prerequisite to salvation. Thank God for that or none of us would make it.

All one has to do is believe God's testimony about Jesus Christ. It's not a matter of understanding the trinity or any other mystery. It's a matter of believing God. To deny Christ is to call God a liar.

Romans 10: 9-13 sums it up. " If you declare with your mouth, "Jesus is Lord," and believe in your heart that God raised him from the dead, you will be saved. For it is with your heart that you believe and are justified, and it is with your mouth that you profess your faith and are saved. As Scripture says, "Anyone who believes in him will never be put to shame." For there is no difference between Jew and Gentile—the same Lord is Lord of all and richly blesses all who call on him, for, "Everyone who calls on the name of the Lord will be saved." There are specific instructions on how to proceed from that point.

UFO's AND ALIENS

Something is out there. It's all speculation at this point but I do believe all will be revealed eventually. I have personally witnessed strange objects in the sky that defy explanation. At 12 years of age, I watched a controlled object move silently across the sky just above my head. It came from near the woods, passed overhead, and curved out over an open field. I was fixed on it until it disappeared. It was 1962 which was a long time before advent of miniature drones appearing in common use. I've wondered if it was an advanced military drone or an actual alien craft. I'll likely never know this side of eternity.

More and more people report on strange sightings today than ever before. If it is discovered that aliens do exist, it could pose serious questions for people of faith. We should not be alarmed.

If an alien race appears on scene and offers explanations without the benefit of the Creator, God, and the gospel of Jesus Christ, they will be easily identified as wicked tools of

Satan. The Bible warns us not to accept anyone who preaches a different gospel. 2 Corinthians 11: 3-4 "But I am afraid that just as Eve was deceived by the serpent's cunning, your minds may somehow be led astray from your sincere and pure devotion to Christ. For if someone comes to you and preaches a Jesus other than the Jesus we preached, or if you receive a different spirit from the Spirit you received, or a different gospel from the one you accepted, you put up with it easily enough." While the world at large may worship at their feet and gleefully adopt alien explanations for things mysterious, the person who knows their God will hold fast to faith.

It's possible that strange objects in the sky are top secret military aircraft or those operated by fallen angels who were thrown down to earth with Satan. They are presently restrained from making overt contact with humanity at large but are waiting for the day when they are released. Revelation 20: verses 7-8 "When the thousand years are over, Satan will be released from his prison and will go out to deceive the nations in the four corners of the earth." Pray for the generation that falls prey to such wicked deception.

Then again, are they time travelers observing humanity's past? Since God can do anything, we must assume that he can permit citizens of his Kingdom to do it, howbeit with restrictions. It's still future to us but it's our past from there. We can't know for sure, but we can speculate that God has allowed saints to review human history in real time. It's fun to think about.

MULTIPLE UNIVERSES

They are thought to exist in several theories, particularly the String Theory. The Time Particle theory says that only two universes exist excluding the tiny junior universe previously discussed. They mirror each other and will eventually consolidate with fusing of temporal time and heavenly time. Ephesians 1: 10, "to be put into effect when the times reach their fulfillment—to bring unity to all things in heaven and on earth under Christ."

Some theorists suppose that a new universe spirals off every time we make a major decision. Simply explained, I decided to publish the theory of Time Particle. Remembering that anything that can happen in quantum theory will happen, I also decided not to publish this theory, thereby creating a new parallel timeline with different results. The idea appeals to many students of quantum physics but I reject it.

Applying it to reality, it would mean that I would experience different results of both decisions howbeit unaware of the other. In one timeline, this theory of time would become accepted by science as valid and I would become an acclaimed theorist. In another timeline, nobody would bother to read about it. It would mean that if I was in an accident, I would escape death and die at the same time. Each decision made would lead further and further into the experience of all things.

In one timeline I would be a ditch digger, which I have actually been. In another timeline I would be President of the United States, which I have never been. At close of time when all things consolidate, I would have been both ditch digger and President. Furthermore, I would have experienced every possible situation and circumstance, every hurt and joy. I would know all things. I would have accepted Christ and rejected him at the same time.

On the day of judgment I would stand before God as God having all knowledge. That is cause for rejection of the theory. Satan wants to be like God. I'm not walking down that path with him. Isaiah 14:14 quotes Satan. "I will ascend above the heights of the clouds; I will be like the most High."

Not me.

Satan's ultimate plan

Satan means "opponent." According to the Bible, his name is Lucifer. He held a high and lofty position within the Lord's army of hosts. He was appointed guardian over the Garden of Eden. He likely participated in creation week. He had great authority and was close to the throne of God. Something went wrong, either by choice or God's will. Satan became filled with pride because of his beauty and level of authority. He became corrupt and wanted to be like God. He rebelled against the Creator and sparked war in heaven. He was thrown to the earth where he dwells among men, or at least the stars.

Satan took a number of fallen angels with him. He is fully aware of judgment waiting. He is presently doing all he can to pervert the time line and prove God to be a liar. If he can do that, Satan can prove that he is the one who deserves to be worshiped.

In the beginning, Satan lost the war for dominion over humanity. He met his final judgment. His past person is now waging war to change the outcome because he knows the future can be changed. His power is real and extremely dangerous. The greatest tool in his war chest is humanity itself.

Disinformation is universal to discourage people from the Christ Jesus by any name recognition. If Satan can succeed, he reasons, it is people who will usurp the final outcome by their own choice.

Satan devised all manner of schemes after his original failure. All of them discredit God, deny Jesus as Savior, and are intended to draw humanity further and further away from God. He and his agents work within religious communities to twist and lead down a wrong path.

Since his present self can read the Holy Bible, he knew early on second time around that God would send a part of himself, Jesus, to the cross in order to reconcile humanity back to himself. Satan tried to jump ahead of that event by creating stories of other 'gods' who sent their son on a similar mission. He uses that, and other things, to place doubt about the real Son of God and the cross. Satan is a clever devil in many areas, but limited in what he can do.

His best hope is to unite humanity under one umbrella of peace with God. That way, he can claim that man does not need God. Hence he would prove God to be a liar and not worthy of worship.

The Tower of Babel story in the Bible sets the stage for what is going on today. A one world government without God is the plan. Human operatives are the key to it's success. It faces many obstacles, but is within reach. Satan's problem is thus: God is supreme. He will not allow Satan to completely succeed.

God has made provision for people of faith. Often utilized clairvoyants, psychics, and occult practitioners have no part in conveying the revelation of God; all they can provide are parts of truth and misleading conclusions. I am referring to the real thing, not parlor tricks.

Deuteronomy 18: 9-14 "When you enter the land the Lord your God is giving you, do not learn to imitate the detestable ways of the nations there. Let no one be found among you who sacrifices their son or daughter in the fire, who practices divination or sorcery, interprets omens, engages in witchcraft, or casts spells, or who is a medium or spiritist or who consults the dead. Anyone who does these things is detestable to the Lord; because of these same detestable practices the Lord your God will drive out those nations before you."

The occult is not all inclusive of devilish schemes but is part of it. Otherwise, the true God spirit-filled receiver is protected and sees what God allows the receiver to see or hear in proper context. It is possible to hear from God through dreams, visions, words of knowledge but anything God reveals will always line up to his character and nature gleaned through the holy written word, the Bible. Satan cleverly uses bits of truth to lure the unstudied away from it. The Bible warns us to test the spirit. We must wrap ourselves in Christ because he is able to shield us from those dangerous, invading entities.

God interacts with us and prophecy is one of the gifts of the Spirit. Romans 12:6-7 "We have different gifts, according to the grace given to each of us. If your gift is prophesying, then prophesy in accordance with your faith; if it is serving, then serve; if it is teaching, then teach."

1 Corinthians 12: 4-11 "There are different kinds of gifts, but the same Spirit distributes them. There are different kinds of service, but the same Lord. There are different kinds of

working, but in all of them and in everyone it is the same God at work. Now to each one the manifestation of the Spirit is given for the common good. To one there is given through the Spirit a message of wisdom, to another a message of knowledge by means of the same Spirit, to another faith by the same Spirit, to another gifts of healing by that one Spirit, to another miraculous powers, to another prophecy, to another distinguishing between spirits, to another speaking in different kinds of tongues, and to still another the interpretation of tongues. All these are the work of one and the same Spirit, and he distributes them to each one, just as he determines."

Ephesians 4:11-13, "And He Himself gave some to be apostles, some prophets, some evangelists, and some pastors and teachers, for the equipping of the saints for the work of ministry, for the edifying of the body of Christ, till we all come to the unity of the faith and of the knowledge of the Son of God, to a perfect man, to the measure of the stature of the fullness of Christ;"

Note that it says, "till we all come to the unity of faith…and to the measure of the fullness of Christ." We are far from that in this present timeline. That means that the appointment of gifts and office are active and available to us. 1 Thessalonians 5:20-21 "Do not treat prophecies with contempt. Test everything. Hold on to the good."

WHO OR WHAT CREATED GOD?

This is the last question posed in this thesis about the Time Particle and things mysterious. Mysterious, it is. It can't be ignored because somebody will ask the question.

I would be a fool trying to answer it with limited knowledge of a sovereign creator but the question looms large for scores of humanity longing for an answer.

Absence of a satisfactory reply is cause for many not believing in a Creator God. They can't believe in God because they can't see him or touch him. Yet they believe in numerous other things intangible.

Hopefully, the theory of time shores support for existence of God. He created it. He is in it. It bears witness to his existence. Regretfully, it doesn't explain how God came to be God but it doesn't matter. He is who he is whether anyone likes it or not.

A personal encounter is required to believe in a sovereign creator. I came to total faith through interaction with him. That will shape up differently for all of us but he is active in our lives past, present, and future.

There are many imitators of God and Christ so one is benefited with a working knowledge of who he is. The Bible is among the oldest manuscripts in the world and testifies to the nature and character of God. Everything must line up to that nature and character or it's not him.

I encourage anyone asking the question to take it directly to God and challenge him to make himself known. One can start a process of entanglement with Jesus Christ through the work of the Holy Spirit with a simple question – "God, is it true? Are you really there? If you are, I want to know you. Look at me, Lord. Don't look away until you have saved me."

God is sovereign. He heard the challenge when presented in the first timeline, and he hears it now. A sincere person should get ready for change. It's coming.

The Time Particle Theory explains how a sovereign God knows the future and how mysterious things work. It's not a new thing. It's always been there. We're just now seeing it.

If I have failed to properly explain parts of the theory understandably, this summary is for the sake of clarity, if that's possible.

Nothing existed this side of heaven until time arrived on the scene. Everything created in our universe was created in time. We exist in time. We have an invisible twin universe. They are identical houses because of the Time Particle's DNA. What happens in one happens in the other. They are separate but both are the same. They are connected to the point of origin by an invisible portal.

Time Band is another name for the universe. We live in it. Everything we know exists in it. The Time Particle spewed dark matter space outward thereby creating the known universe. The Band's DNA compels it to orbit in a perfect, continuous spiraling vortex that excites dark matter. That motion combined with gravitational waves creates ripples in it. The energy pushes dark matter against the Band wall and extends it making more room for time. Bands are speeding up with each orbit. The point of origin and other points past get further and further away from the present but they entangle each other. Time moves on but it leaves a permanent record behind itself. It is God's memory book.

Temporal time within the band has its own set of rules that govern our lives. All things within the Band react to those specific rules.

Time bands interact. Both float within the boundary of the atomic house and cannot exist outside of it. They crisscross and bleed over into each other at close intervals. Each second of the time band is full to include the past, present, and future. We live in all three states and each state is as real as the other. We live presently, we are living in the past, and we are living in the future. Band entanglement results in paranormal encounters such as precognitive dreams, visions, and other spooky things. God spoke and speaks to people through time by sending messages back to humanity through prophets because we live in his past.

Only God can change the past and alter future events happened. He is the Great Observer watching us. Our present is his past. That's why we should pray for reset of our future events and direct the course of them through prayer. Satan is trying to alter the future because he knows it can be altered. Our prayers can affect other people's circumstances as well as our own.

~

Thank you for reviewing this theory of time and things mysterious. Please consider leaving a review on the Amazon website for "The Time Particle" by yours truly. Let me know what you think about it. Serious input is sincerely appreciated. God's best always.

Terry McIntosh

www.ingramcontent.com/pod-product-compliance
Lightning Source LLC
Chambersburg PA
CBHW021452210526
45463CB00002B/753